Giant Pandas and Me:
Ten Years Of Discovery

By

Keith E. Jones

No part of this publication may be reproduced, stored in a retrieval system, transmitted in any form or by any means, electronic, digital, mechanical, photocopying, recording, or otherwise without written permission of the copyright holder.

Copyright 2013 Keith E Jones

All rights reserved

ISBN-13: 978-1492306856

ISBN-10: 1492306851

First Edition

Printed in the USA

Table of Contents

Introduction .... Page 7
The Giant Pandas and Me

Chapter One: .... Page 15
Giant Panda Basics

Chapter Two: .... Page 19
A Plan To Save The Giant Pandas

Chapter Three: .... Page 27
Developing a Successful Breeding Program

Chapter Four: .... Page 43
Releasing Pandas Into the Wild

Chapter Five: .... Page 49
Dangers to Wild Pandas

Chapter Six: .... Page 53
Tracking Pandas In the Wild

Chapter Seven .... Page 55

About the Author............................Page 63

Reading ....Page 66

Introduction

The Giant Pandas and Me

My introduction to the giant pandas came in April 2004 when I embarked upon my first trip to the Wolong Panda Preserve. What an unbelievable experience that was for me. That idyllic period ended for me when the 8.0 magnitude earthquake destroyed the Wolong Panda Preserve on May 12, 2008. Since then my volunteer tours have gone to the Bifengxia Preserve.

As I entered Wolong that first day pandas were seemingly waiting around every corner of the panda preserve. The next five years were what I now think of as my special years of the panda. During that five year period I was at the Wolong Panda Preserve almost every month for a week or more, beginning in April and continuing until November. I worked as the group leader for tour groups I organized to do volunteer work at the Panda Preserve.

During this 4 to 5 year period I was at Wolong so much of the time that the staff, who had great difficulty pronouncing Keith, gave me the nickname, Lao Ki. My Chinese speaking friends laughed when they heard that. They told me only half jokingly that the name means Old Man Keith.

The Wolong Panda Preserve was a special place for me like no other on our planet. The Preserve facility is nestled in a narrow river valley, trapped between the steeply rising mountains on two sides. The mountains and the Preserve are frequently shrouded in a misty cloud layer that adds a mystical magic to the feeling of the facility. The Preserve was and is a relatively small facility.

This closeness of enclosures was the topic of a research paper in 2006 that criticized the size and how close the enclosures are one to the other. The paper postulated the close proximity of one panda to another could allow disease to spread like wildfire through the entire facility.

This critical research paper threw the Wolong staff into a worried frenzy of anti-bacterial counter measures. Should some epidemic have passed through the enclosures from one panda to the next, then the administrative staff would have lost great face with the top China leaders.

After publication of the paper the Chinese government set into motion a series of plans designed to change the focus of work performed at the Wolong Preserve. The goal was to transition to a pure research and breeding facility unimpeded by the demands of catering to hundreds of tourists each day. Construction on new panda facilities was started in several locations away from the Wolong facility.

The idea was to eliminate the large flow of tourists through the Wolong Panda Preserve facility by diverting these bus loads of tourists to the new facilities. Had the earthquake not destroyed much of the Wolong Preserve, then the tourism flow would have eventually been diverted 100% to the new facilities.

The purpose of this tourist diversion was to keep the hordes of tourists from carrying some disease to the pandas at Wolong. I heard of those rumors in late 2006, as the plans were first being set in motion.

For me the closeness of everything at the Wolong facility wasn't a cause for concern. Rather it made Wolong a wonderful place for me to just hang out. While my tour group members were performing their volunteering activities I was free to wander through the entire Wolong facility. There were only a couple highly restricted areas that were off limits to me. The veterinary hospital and the baby panda nursery were those two areas.

The Wolong Panda Preserve became like a personal meditation ground for me. I climbed to the highest hillsides surrounding the preserve. There I would sit and meditate for hours on end. Sometimes the mountainsides were shrouded in fog. My fleece coat would glisten in the refracted light as the dew drops formed on my outerwear.

I had plenty of time to myself while my volunteer groups worked with their assigned panda keepers. Alone, I roamed the mountains and forests around Wolong. I was a solitary foreigner in search of wildlife and solitude. During my wanderings I discovered two wild red pandas who lived in the area near the Preserve, once I sighted a small clouded leopard and several times some wild boar.

In the 5 years I roamed the Wolong Preserve I never saw another person climb this hanging walkway. The landing at the top became my personal meditation site.

When the terrible earthquake struck in 2008 the Wolong Panda Preserve was devastated. I had a tour group at the Preserve two weeks before the earthquake and another group scheduled to visit there just one week after the quake struck.

I read every report out of the earthquake zone with a heavy heart. I feared I would read that some of my friends had died. Thankfully they were all spared.

After the earthquake the Bifengxia Panda Preserve became home to most of the pandas that had lived at Wolong. The majority of the staff from Wolong were also relocated to Bifengxia. In an effort to keep the very successful Wolong breeding program going, the Chinese government moved staff, administrators and pandas to the much larger facility.

This move gave the staff access to a real city, Ya'an. The Wolong staff moved to an easier living environment. The pandas were placed in much larger, far more widespread enclosures. The worries about an epidemic moving through the entire facility diminished to the normal concern levels that all zoos and animal keeping facilities share.

The earthquake and subsequent move to Bifengxia ended my 5 years of panda bliss. While the Bifengxia facility is nicer, newer, larger and in every way a better facility for the staff and for the pandas, for me it just did not have the magic of Wolong.

I found myself suddenly on the outside. At Bifenxia I was no longer Lao Ki, but just some guy roaming around the very large facility in search of his old friends. The pandas were still accessible for me, but the freedom and access I had been given at Wolong was absent (and still is) when I was at Bifengxia.

Sure many of my old friends from Wolong are there at Bifengxia. The facility is large and it is a great place to house lots of pandas. It is a great place for tourists to visit. But that special connection I had at Wolong is gone now.

Now I enjoy roaming the old historical sections of the City of Ya'an. It is an interesting town to be based in. There is much more for the one time tourist to do and to see in Ya'an and Bifengxia, when compared to Wolong.

I mention this comparison between the two locations because as I wrote this book I found my personal bias toward Wolong creeping into my choices of photos and into how I worded some things. Like your first broken hearted teenage love affair, my romance with Wolong will always be the one that I think about with a heavy heart full of nostalgia for what once was, but can never be again.

For the record I think that the facility at Bifengxia is a superior location for the care and breeding of captive pandas. It provides the same or a superior experience for the one time visitor.

My panda volunteering tours continue, but now I seldom wander through the facility at Bifengxia as I did at Wolong. The Wolong staff and my friends from there still talk to me, but the special connection between me and Wolong is missing at Bifengxia.

Perhaps that's why now I only get to the Bifengxia Panda Preserve a couple times each year. Instead of the 50 or 100 days a year that I enjoyed at Wolong, I only visit Bifengxia 5 or 10 days each year.

For you readers who are panda enthusiasts, the experience of doing a few days of volunteering work at the Bifengxia Panda Preserve would most likely be the most memorable week you will ever enjoy while on vacation. It is an experience that anyone fascinated with pandas should have at least once.

Now on a different subject, as you read this book you may question why I don't write more about the dark side of the pandas. Why do I mostly ignore issues such as the money making machine the Chinese government has set up to lease pandas to various zoos around the world? Why do I choose to ignore the issue of captive pandas being subjected to close contact with dozens of tourists every day?

These and other negative topics about China and the pandas have been thoroughly beaten to death on hundreds of internet websites and in several recent book offerings. At the end of the book you will find a bibliography that gives some additional suggested reading. Some of those books delve heavily into these and other negative aspects of the Chinese Panda Conservation Plan. If you want to learn more about the negative side of panda politics, you can find information in those books.

In all my writing about wildlife and about foreign culture I choose to see and to reflect upon the brighter side of life. I made a conscious decision more than 20 years ago, as I began my career of working with animals and travel that I would present the positive and leave as much of the negative to someone else to write and talk about.

I find plenty of good to write about. In my writing about whales I don't discuss whaling nor do I offer photos of dead whales. In this book I mostly ignore poaching, trapping and abuse of pandas and you won't find photos in this book depicting those things.

I hope you enjoy what I have to tell you about the giant pandas.

You're welcome to contact me about this book, about traveling with me or with questions about pandas or China.

Email:
keith@greywhale.com
rowman1998@yahoo.com

Skype ID: bajajones

Website: **www.bajajonesadventures.com**

If success is measured in numbers alone, then the China Conservation Captive Breeding Plan is a big success. Now averaging more than 30 babies born annually, the program has fully achieved the goal of learning how to successfully breed pandas in captivity. When started the goal was to have 300 captive pandas from the breeding program. That goal has been passed and a new goal of 500 put in place.

Chapter One
Giant Panda Basics

Kingdom: Animalia
Phylum: Chordata
Class: Mammalia
Order: Carnivora
Family: Ursidae
Genus: Ailuropoda
species: melanoleuca

Classification: DNA and other scientific studies currently classify the Giant Panda in the bear family. So the old name of Giant Panda Bear is still correct today. The pandas are no longer considered related to raccoons.

Size: Standing height about 4 feet. Overall length when lying flat on their back is about 4 to 6 feet maximum. Their height when on all four legs is about 3 feet tall.

Weight: Males are larger than the females with a maximum weight of around 300 pounds, perhaps a touch more if they are overfed in captivity. Females generally weigh less, around 150 to 250 pounds, again perhaps a bit heavier in rare cases.

Color: Black and white, along with zebras, pandas have perhaps the most recognizable color pattern of any animal.

Age: In captivity the normal age span is around 25 years, with the oldest known pandas living to 37 years. Nobody knows for sure how long pandas live in the wild, but scientists speculate a life span of 20 to 25 years in the wild.

The panda who is thought to have lived the longest was named Dudu. Dudu was born in 1962 and died in 1999 at the age of 37 at the Wuhan, Chengdu Zoo.

Reproduction: Mating March to May. Gestation period about 150 days, but actually a wide range from 90 to 180 days

Habitat: The pandas are found in rugged mountains where a combination of hardwood and conifer trees shade a dense undergrowth of bamboo. They migrate up and down the mountain elevations as the weather changes with the seasons. Found at elevations as low as 1,000 feet in the winter, summer might find a giant panda at an elevation of 10,000 feet in elevations close to the tree line.

The climate in these forests is wet and generally cool. An overlying characteristic is damp moist air. This moisture helps promote the dense bamboo growth that is key to the giant panda's survival.

Population: The 2003/04 panda census in China estimated 1,600 Pandas in the wild. Research in 2006 using DNA gives hope that the real number might be as much as 3,000. The current ten year census began in October, 2012 and is expected to finish October 2013.

Food: We all know that pandas eat bamboo. In the wild they also eat some other forms of vegetation and fruit if they can find them. I have seen pandas eating grass on occasion. Pandas are omnivores and as such they will eat birds eggs and small rodents if they happen to catch one. But bamboo still makes up as much as 99% or more of their normal daily diet.

You have only to see panda poop in the wild to realize bamboo is the main food source for the pandas. Bamboo is a food that their digestive system is not very good at processing.

A small pile of panda poop in the wild will be bamboo green in color and have a texture like woven or pleated bamboo mats. It seems the bamboo passes through the Giant Panda's digestive system with a good percentage of the vegetation not processed into energy. That's why pandas eat 12 to 18 hours a day.

In captive conditions the pandas receive other food along with the bamboo. The special panda bread is just one nutritious food used to supplement their bamboo diet. Apples are a panda favorite. They get fed carrots too, but those are not such a treat for them.

Chapter Two

A Plan To Save The Giant Pandas

After 10 years and millions of dollars spent in well intentioned but at times half hearted studies and field research, in 1988 the World Wildlife Federation presented a plan to the Chinese government. The research leading up to the writing and presentation of the WWF panda conservation plan to the Chinese government is the topic best left to another book. This was a big plan for saving one of the most recognizable species on the planet from extinction.

Suffice it to say here that in the 1970s to 1990s China was a very different place than it is today. Foreigners were not so common. Then President Nixon visited China in 1972. Thus opening the first meaningful non-warlike dialogue between our two nations for many years.

Cultural differences led to misunderstandings between foreign and Chinese scientists. Those cultural differences are still in play today, although not always as blatantly obvious as in those early days of panda research.

The WWF plan with nearly 10 years of field research behind it, was shuttled from one bureaucracy to another. Then the tragedy at Tiananmen Square happened and the WWF plan for panda salvation just floated in limbo, pushed into the background by the horror of that week.

In 1991 the China Ministry of Forestry (which administers the giant panda population) finally introduced its own panda conservation plan. Loosely based upon the WWF recommendations, but seriously modified from that very detailed proposal, this Chinese 5 year plan for the years 1991 to 1995 had a budget of 51 million dollars attached to it. This plan was eventually ratified in January of 1992.

For me, 50 million dollars seems a small amount when put into the perspective of the trillion dollar Chinese economy. For the Chinese at that time it was an extravagantly huge sum of money to spend on any wildlife conservation plan. Initially the plan only received partial funding.

The major elements of this conservation plan included developing 14 Forest Panda Preserves that would be set aside as a natural habitat for the pandas. The original WWF plan called for migration corridors connecting 35 preserves so as to allow for a viable and energetic population of pandas, not simply for separate and isolated populations of pandas.

The main elements of the Panda conservation plan are listed below. Alongside each element of the conservation plan I have given that goal a letter grade, like students would receive at school in the USA. A is excellent, B is good, C is barely satisfactory, D is not good and F is for failure. The grade I give them is based upon my personal belief in how successful the Chinese have been to date at achieving these goals. My opinion is just that, opinion. Ask a dozen other knowledgeable panda people and you will likely get 12 different answers:

Elements of the Conservation Plan with letter grades

- **Reduction of human activities in the panda habitat**
  At best I give this a C- grade.

- **Removal of human settlements.**
  The Chinese government can be harsh when it wants to accomplish something. But it has moved slowly in trying to accomplish this task. The grade for this falls between C- and D+.

- **Modification of forestry operations**
  In some areas the goal has been achieved, while in other areas the old status quo continues unabated. Give this a C grade.

- **Control of poaching.**
  From what I have been able to learn, today, 2013 this would rate perhaps an A- or even an A. Strong penalties and a national love of the panda have reduced poaching to very low levels.

- **Rehabilitation of habitat.**
  Let's give this a strong C. There are actually areas in the mountains where bamboo has been planted and where conifers are growing again.

- **Management of bamboo habitat**
  This really goes right along with controlling forestry and rehabilitating habitat. But here I give the Chinese a strong B grade. Before the 2008 earthquake more local people around the panda preserves were moving into growing and harvesting bamboo.

- **Extension of the panda Reserve system.**
  Perhaps a D or even an F grade for this effort. Although there are now 60 designated nature reserves and breeding centers in China I think the large wildlife preserves that were proposed in the WWF plan in the 1990s have not been fully realized. Most importantly for the future of the panda species the connecting wildlife corridors haven't been developed or protected.

- **Achieving out breeding between panda populations**
  Give this a big ? question mark. I think the existing wild pandas are not breeding from one isolated area to another. I'm not sure how to achieve out breeding between these isolated population unless they 1. Capture and relocate female pandas or 2. Release captive bred pandas into these wild populations. Neither of these activities is happening.

- **Maintaining a captive population**
  This is the major success point of the panda recovery plan. Give this an A+. There are now too many captive pandas being born each year.

- **Release of captive-born pandas into the wild**
  Still in the very earliest stages of development. Stalled by fear of failure. This is the next major goal the Chinese are concentrating on. But at this time I can only give the re-introduction program a D- with hopes that in the coming years this letter grade will become an A+.

These targeted tasks have been underway since the mid-1990s with various levels of success. I want to avoid the politics of pandas as much as possible in what I discuss in this book. But it is impossible to talk about pandas in depth without getting into some form of politics.

Most of the conservation plan goals have been only marginally successful. Worse, on some of these important conservation goals there has been no positive progress.

The current panda census now underway will most likely show that the population in the wild has grown or stayed the same size as ten years ago. This might be considered a minor victory for the panda conservation program. Or it might mean that with more modern census techniques the count is more accurate, but not pertinent when compared to the census of ten and twenty years past.

A real major victory will be when the final element of the conservation plan (release into the wild) is running as successfully as the breeding program is running now. That final goal of releasing of captive pandas back into the wild is in my opinion the key to panda species survival. This topic will be discussed in detail in Chapter four.

What good is panda poop?

Did you ever wonder how researchers go about counting all the pandas in a country the size of China. Obviously it is not an easy task. Here is an animal that almost nobody living today has seen in the wild. The number of people who have even seen one wild panda is incredibly low.

Are these panda researchers so good at tracking that they were able to locate and count 1,600 pandas when they did the last census? How can they be certain they counted all of them?

The Los Angeles Times printed an article written by their Beijing bureau chief, Barbara Demick. In the article Ms. Demick quotes a Chinese lead researcher involved in the 2012/2013 panda census.

"To be honest, I've been working in these mountains (in the Sichuan Province) for 20 years and I've never seen a panda in the wild," says Dai Bo, 43, a wildlife biologist with China's Forestry Ministry.

During the current census for more than a year the researchers will go out daily into the mountains. There they will walk predetermined map quadrants collecting samples of panda poop. These samples then get taken back to the labs where each sample is carefully recorded on a map with GPS coordinates of where it was collected.

The panda poop samples are tested and visually inspected one by one. Size and consistency of the bamboo pieces contained in the sample is noted. This information can be used to gauge the age and some health issues of a panda.

The sample is also DNA identified. This allows the researchers to tell how many different pandas produced the various droppings they are working with and exactly where each panda traveled during its daily wanderings.

Chinese tracker shows me fresh panda spoor. Lucky man that I am, later that day we saw the black and white butt of a wild panda as it slipped away from us through the damp, dense umbrella bamboo in the Qinling Mountains at an elevation of 6,200 feet, July 17, 2006.

More than 100 researchers, working for more than a year will count, tabulate and eventually issue an official census. This is a massive undertaking whose results will tell us a lot about how the Chinese Panda Conservation program is working out.

This census might evaluate tons of panda poop during the full census year and that's what panda poop is good for.

As I write this it is May, 2013 and the census figures are not in yet. I am so interested to see what the results will be.

Chapter Three
Developing a successful breeding program

Most of us who are fascinated with giant pandas know that they are not prolific breeders. Unlike rabbits or mice, pandas are normally solitary in the wild and don't produce hordes of little pandas every year. But given proper habitat and natural conditions the pandas survived for thousands of years, so while not prolific breeders, they managed for a long time to get it right.

Female pandas living in the wild might give birth once every two to three years beginning at age three or four. The likely continue to conceive until around 15 or 20 years old. This means any wild female panda could successfully raise 4 to 6 cubs during her breeding lifespan.

The life style of the giant panda contributes to the poor birth rate. Normally solitary, migratory in the sense that they move up and down the mountains with the seasons, not very social even during the mating season and only capable of conceiving when their physical condition is just right; there are just so many obstacles in the way of a successful mating it is a miracle any baby pandas are ever born.

The pandas occupy a wide territory constantly on the move across the mountainsides. For the longest time it was thought that only one panda would occupy a territory several miles in area. But recent DNA studies of panda poop proved those earlier theories wrong. It is now accepted as fact that several pandas can live in the same overlapping territories.

On occasions these pandas might even come together and socialize. Although I think this is wishful thinking on the part of some researchers. My extensive experience with pandas both in observing them in captivity and in the wild, leads me to believe that after they get to be 3 or 4 years of age they are not very social at all. In the breeding centers it is dangerous to house more than one or two pandas older than 4 years in the same enclosure.

When the Chinese government began a last ditch effort to save the pandas, what was known about them was mostly folklore and hunting lore passed from local people to researchers. Scientists didn't know enough to manage a successful breeding program.

The learning process was slow going at first. The failures were frustrating and emotionally wearing on both the scientists and panda keepers, as they lost one baby panda after another.

Li-Li, was one of the first giant panda females at the Wolong Panda Preserve. She was brought there from the Beijing Zoo in 1980. She gave birth to the very first panda cub ever born at the Wolong Panda Preserve in 1986. The cub, named Lan Tian, only lived until 1990.

Even then the Wolong Preserve was the acknowledged leader amongst all panda centers for the care and welfare of pandas. But they still had so much to learn.

The things that they didn't know when the breeding program started would fill many books. In this chapter I will touch on some of the highlights that stand out for me in the development of what must now considered to be one of the world's most successful wild animal breeding programs.

Why is this program different than say the pronghorn antelope breeding program that the Vizcaino Biosphere Reserve in Baja California Sur, Mexico has developed? Simply by the complexity and difficulty that the researchers faced in understanding and then implementing a breeding program.

While the wild Pronghorn Antelope population in Baja is almost extinct because of hunting and competitive grazing by cattle, the breeding program was quickly successful, once funding became available. Put several antelope of opposite sexes together in a really big enclosure, feed them well, protect them from coyotes and cougars and come Spring a new crop of baby antelope are sure to be born. Most of these newborn antelope, when reared in captivity, will grow to be healthy adults.

Do that same thing with several giant pandas and you will have complete failure. No baby pandas will be born! Maybe there will even be fewer pandas if one of the females takes a dislike to some horny male and attacks him.

Those were some of the challenges that the Chinese giant panda breeding program faced when it started. An almost complete lack of animal husbandry skills with pandas, no real scientific understanding of how to get them to mate successfully and a species that by nature is almost shrewish and hermit-like. Even during estrus the female isn't really receptive to mating a good percentage of the time.

Here's a story from the early days of panda breeding in China.

The headline of an ad placed by the Beijing Zoo in the Beijing Youth Daily newspaper read:

"Panda Seeks Mama Dog"

"Lele, the Beijing Zoo panda who had twins on August 3, does not have enough milk to feed both of her babies. Because she cannot care for two cubs, she chose to nurture one and abandon the other." So read the news article and ad.

Lele's zookeeper said that this is a "normal and instinctive" behavior for panda mothers who don't produce enough milk.

What can a zookeeper do in the time of a panda milk shortage? Ask a dog for help! Because dog milk is very similar to panda milk, zoos can use dog milk to feed a baby panda.

"The response was very enthusiastic," said a zookeeper. The rejected panda baby soon had a surrogate mother dog.

This took place in the year 2000 when panda breeding was still a hit and miss proposition. That year 8 baby pandas were born in captivity in China. Compare that to the 30 or 40 that might be born this year.

How did the Chinese develop such a successful program in the face of almost overwhelming obstacles? Below I will describe some of the innovations and successes the program has had.

Most of all, I believe that the giant pandas were saved because every staff member at the Wolong Panda Preserve was unusually dedicated and focused on their task. Through luck, good fortune and skillful hiring practices the Preserve quickly built a unique staff that worked tirelessly for years without complaint and without losing focus on their goal. I am the biggest fan of each and every one of them. I'm proud to have some of them as my friend.

The researcher's ability to focus on minute details and to study some detail for hours on end without losing concentration was a start in the process. The skill and luck in hiring that contributed to developing a staff of panda keepers at Wolong who love those pandas as if they are their own children, was instrumental in the success of the Wolong breeding program.

The money donated by foreign non-profits organizations such as the World Wildlife Federation (WWF) and Pandas International was truly helpful, especially in developing the tourism aspects of the breeding program, thus seeding a renewable cash resource to fund more research and to help purchase additional, expensive scientific and medical equipment.

Those few of you who have been to Wolong or to the new breeding center at Bifengxia know that the Panda breeding centers are distant from cities and relatively isolated. Cultural activities and social events are almost non-existent. The staff of the center must be willing to live in isolation throughout the winter. Roads become almost impassable for days or weeks on end as winter storms pile up ice and snow, clogging the highways and side roads.

The work conditions are difficult in the winter. It is not easy to work outside in sub-zero temperatures, with snow or icy rain falling. Cleaning cages, dragging 40 kilo bundles of fresh bamboo to the pandas, shoveling panda poop and washing down a dirty enclosure are more like torture than work when performed in freezing rain. The staff endures these difficult conditions stoically and without complaint.

Being able to eat, drink and work with the staff for weeks on end, I learned about many of the techniques they have developed at Wolong for breeding pandas. Every technique learned was another baby step toward success. After 20 years and dozens of heart breaking failures all those baby steps now add up to one giant panda breeding success. Those failures, those tiny successes, and every new technique developed, moved the program forward to where it is today.

The advances in animal husbandry have been enormous and are a key factor in raising the survival rate of babies.

1. **Clearing the baby panda's throat** of milk so they do not suffocate. Many tiny babies were lost from this in the early days of the breeding program.

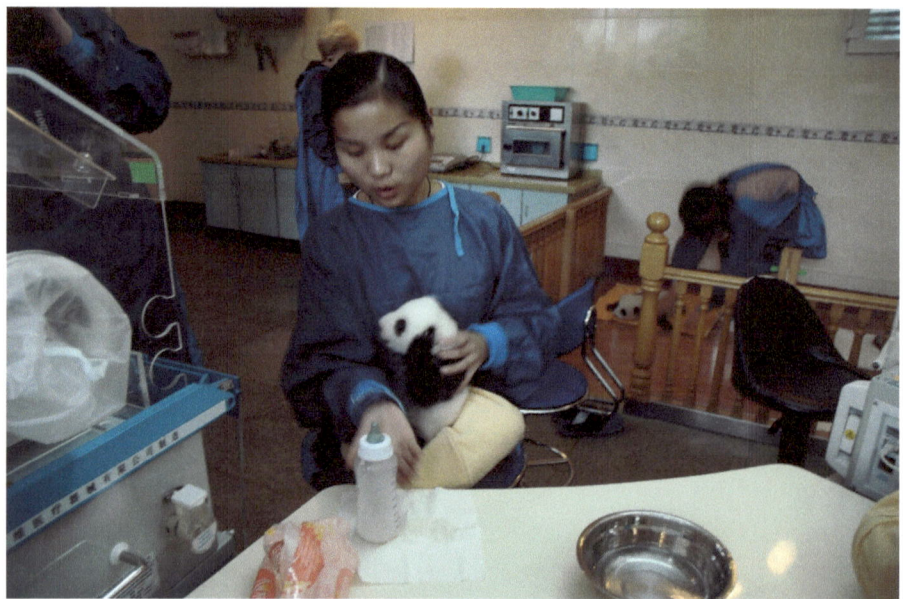

2. **Burping the baby panda** promotes digestion and helps prevent unexpected "spitting up" that can suffocate the baby.

3. **Using human baby incubators** to provide warmth when the baby panda is in the care of the nursery workers.

4. **Separating twins from the mother** has saved many baby pandas. In the wild only one of the twins would survive. The staff rotates which of the babies is left in the care of the mother. When not with the mother the other twin spends time in an incubator and is bottle fed using formula. The mothers accept the twin back without a problem.

5. **Mother pandas undergo an interesting change** in disposition immediately after giving birth. Prior to birth the mothers may be aggressive and fiercely dangerous to both panda keepers and to male pandas. They can be difficult to get close to without protection. After giving birth they become docile. Mothers spend an inordinate amount of time just sitting and laying around. Separating a baby panda from the mother, for instance when trading one twin for the sibling, requires great patience on the part of the panda keeper

The mother may be reluctant to release the baby. She might keep that baby close to her and within the grasp of her huge paws. If the mother doesn't want to release the baby, the keeper cannot simply rip it from the mothers grasp. The risk is too great.

So for hours on end the patient panda keeper will repeatedly approach the mother and try to remove the baby. They will use fruit, panda bread and other goodies to tempt the mother. Eventually, sometimes after many hours the baby is released and the panda keeper then rushes the baby from the mother to the nursery. Once at the nursery, the baby is measured, weighed and given a clean bill of health before receiving a bottle of formula, a burping and then a warm bed in the tiny incubator.

6. **Care of the mothers has improved.** Every year pregnant mothers are given extra nutritious food to promote the healthiest baby possible. A delicacy for the Giant Panda, not normally provided, are fresh sprouts of the Giant Bamboo imported to the mountains from the south of China. These giant bamboo sprouts are tender and tasty and the pandas really love them. Of course apples and panda cake are also provided in larger quantities than normal.

Pre-natal care at Wolong and Bifengxia includes sonograms of the mother, showing the panda fetus. It is interesting to watch a veterinarian and the panda keepers take the sonograms. I have seen photos these sonograms, but to my uneducated eye, they are just a multi-colored free form image.

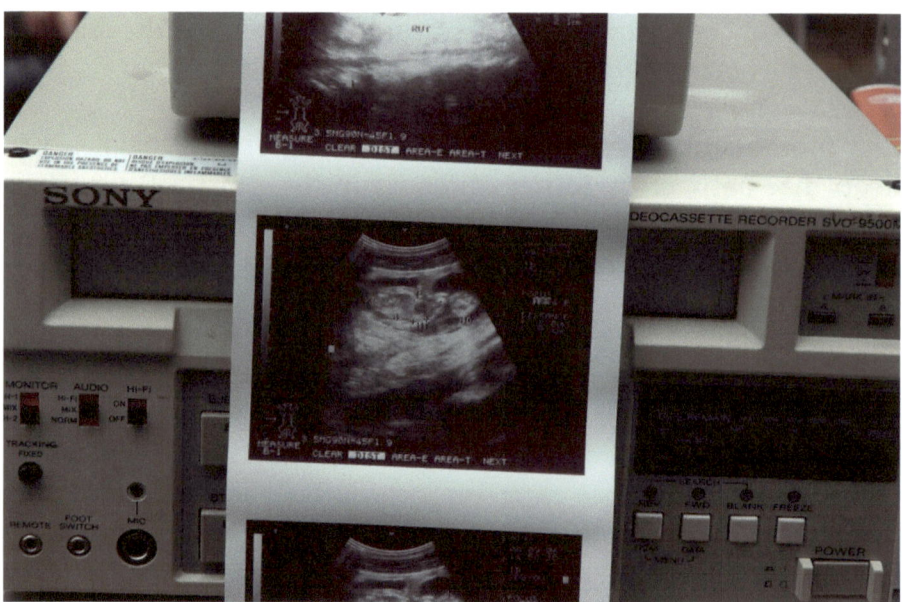

7. **A sonogram** cannot be taken unless the panda has been trained to lay on her back just a certain way. To this end, a series of cages have been developed. These cages are used for transporting the pandas around the Center. They are also used when the vets must work on the Pandas. It is not only performing a sonogram that requires this type of specialized equipment.

To give a panda a health check up requires this same system. If the vet needs to draw blood they will put the panda in one of these cages. Then too, in order to manipulate the male to obtain sperm for artificially inseminating females who don't become pregnant through natural means, the veterinarians must place themselves in very risky positions with their hands and arms inside the cages. Extensive training is required to keep the males from seriously injuring the staff during the delicate procedure. This training is carried out year around and is another important technique developed at Wolong that has helped in so many ways to move the breeding program forward.

8. **Nutrition is a key factor** in panda breeding as well as in increasing the panda life span. A panda is old at 25. The famous Panda bread or Panda cake was developed at Wolong. It is nutritious, tasty to the pandas and healthful if they don't eat too much. (Yes, I have eaten the panda bread and it tastes okay, but it is rough on the human digestive system because of the ground up bamboo).

For the pandas, the bread is too rich in large quantities. The panda keepers must balance between extra nutrition and too much rich food. To this end, every piece of panda bread is carefully weighed for each animal. (Cutting and weighing the bread is one of the tasks that our volunteers frequently help with.)

Here's the "secret" recipe to this tasty cake. The mix consists of corn meal, flour, ground bamboo, carrots, apple, oil and some other secret tasty ingredients all ground or blended to be fine and powder-like in consistency. This is mixed with a little flour and possibly soy milk and then baked in an oven. The resulting "cake" is about 4 to 5" thick and around 10" diameter and looks remarkably like a wheel of cheese when a tray of them are set out to cool after baking.

9. **Strength training is important**. Male pandas must stand on their rear legs for a fairly long time when mating. Their leg and hip strength is not naturally adequate for the male to maintain the necessary position long enough. This is one of the main reasons natural pregnancies in the wild are at such a low rate.

The Center has developed a series of strength and agility exercises that the pandas are put through several times a day. When I was in high school we had a cheer we did at football games, simple but fun to do. The cheerleaders would lead us as we all chanted "Stand up! Sit down! Fight Fight Fight! Then again. . . "Lean to the left, lean to the right, Stand UP, Sit down fight fight fight!"

Well that is almost exactly the exercise we have the male panda perform as his training exercise. Using panda cake as a reward the male panda is made to stand at the front bars to his cage, then move to the left, move to the right, stand up, sit down and then stand up again. This maneuver is performed over and over to strengthen the lower legs and thighs and has led to a much greater natural breeding success rate.

There are some other tricks that were developed at the Wolong Panda Preserve. Simple things such as changing the bamboo given to the panda on a daily basis. The bamboo is trucked in fresh daily. Pandas are great connoisseurs of bamboo. If bamboo is not fresh, the pandas will not eat it. If it is too dry they will not eat it. They can be quite fussy about their bamboo and the preserve staff has learned to accommodate this finicky diet.

Confessions of a bamboo thief

There was a time when uneaten, one or two day old bamboo was not removed from the panda enclosures. The old drying bamboo was left for the pandas to eat or not. Panda keepers just thought that pandas would only eat one or two varieties of bamboo. The world thought that pandas were just too finicky about their diet.

We know better now. Pandas will eat any of some 35 varieties of bamboo. But it must be fresh, like homemade bread is best eaten fresh from the oven. Old bamboo just won't do for these bamboo gourmands.

Around the panda preserves at Wolong and Bifengxia are groves of bamboo. These are grown for landscaping and decorative purposes and are not put there to be used as panda food. The bamboo for pandas is purchased from local farmers who are contracted to grow and harvest fresh bamboo.

But this contracted bamboo comes to the preserves a day after cutting, perhaps two days after being cut. The pandas recognize this difference between fresh bamboo they harvest as they eat, but they learn to accept this harvested bamboo, which is kept covered and moist until it arrives in the morning for delivery by the panda keepers to the panda enclosures.

Several times I helped panda keepers sneak into these ornamental bamboo groves to illicitly harvest tasty, moist fresh young bamboo for some special panda. This might be some panda who was not eating well or who was recovering from one of the frequent stomach disorders the captive pandas suffer.

While my friend the panda keeper would casually slip into the grove of 6 foot or 7 foot high bamboo, her knife hidden in her jacket, I would play lookout. Then quickly as possible the panda keeper would harvest a dozen 4 foot long fresh young bamboo stalks and we would quick walk back to the panda enclosure. The grateful panda would always chow down on that illicit bamboo like it was 20 year old bourbon during prohibition.

The evidence would disappear in 15 minutes. This illicit cutting of the bamboo was considered a minor infraction of the work rules. Something akin to a parking meter violation vs. the more serious crime of drunk driving.

The punishment for this illicit cutting of decorative bamboo would have been for the panda keeper to take on some extra work duty. Work such as working a double day and then night shift during the panda mating season. Or perhaps taking over the work of a sick or vacationing panda keeper while also continuing the care of her own 2 or 3 pandas.

What I have to wonder right now, is if as a self-confessed bamboo thief, when I go for my next visit to Bifengxia in October will I be penalized and given double panda poop scooping duty?

Photo below: The man responsible for catching bamboo thieves. . . . .

Mr. Han, my good friend and the head panda keeper responsible for all of the panda keepers at Wolong until 2008. He has had the same and added responsibility at Bifengxia since the 2008 earthquake.

Combining all the science that has been developed, plus the many learned lessons in animal husbandry and adding in the special tricks that the individual keepers learned and passed along with enthusiasm to the other members of the staff, China has created a most extraordinary and special animal breeding program.

The Sichuan earthquake was a setback. China may have one of the world's healthiest and fastest growing economies, but money is still a precious commodity and hard to come by.

In the years leading up to the 2008 earthquake the Wolong Panda Preserve gradually increased the number of baby pandas born each year. In the year 2000 when the program was in its infancy, 6 baby pandas were born to 4 mother pandas. In the same year 2 more were born at the Beijing Zoo. Bringing the total born in all of China to 8.

The learning curve at the Wolong Panda Preserve was steep. By 2005 the breeding program was a well established success. That year a baker's dozen (13) baby pandas were born at Wolong. In 2006 there were 16 babies born at the Wolong Panda Preserve. In 2007 that number grew to 21. The 2008 earthquake set the breeding program back, but in 2012 there were 20 babies born at Bifengxia and 8 born at the Chengdu Research Facility.

This was the first time in history that so many baby pandas were all together at one time in one location. The date was October 31, 2005. All the major China news services were on hand to capture these same images as I took. They were inside the glass of the nursery while mine was taken from just outside the glass enclosed nursery at Wolong. The photos of these 13 pandas were on every news program almost anywhere in the world in November.

The success of the breeding program is placing a strain on the China's government's ability to provide housing for all the new pandas. The stated goal was 300 captive bred pandas in captivity. With that goal achieved, the Chinese raised the target to 500 captive pandas, because they are reluctant to slow the pace of their breeding program. With the breeding program running at the pace of a high speed train, what comes next?

The answer: Releasing pandas into the wild to bolster the actual wild population of pandas.

Chapter Four

Releasing Pandas Into the Wild

In April of 2006 Xiang Xiang the first panda trained to survive in the wild, was released. Xiang Xiang was a five year old male panda. Sadly, but not unexpectedly Xiang Xiang survived only one year in the wild.

It is incredibly difficult to train animals born in captivity to learn to survive in a dangerous wild habitat. Xiang survived a year in the wild. In the end it appears an accident was the cause of his death.

Researchers speculate that Xiang Xiang was attacked by another male panda or pandas. To escape the attack he climbed a tree. For some reason Xiang fell from the tree and died. He died on December 22, 2006.

This death put the Chinese panda release program into a coma. The cultural situation of "losing face" by being responsible for the death of the panda placed the release program on hold as the administrators worked on a way to minimize the damage to their reputations and careers.

After rethinking the captive panda release program, the new plan consists of 4 phases.

Here is the history leading up to the release of the second panda, Taotao or Tao Tao on October 11, 2012 into the Lizipeng Nature Reserve.

Phase 1 began on July 20, 2010 when two pregnant female pandas were placed into the semi-wild enclosures at the Hetaoping Research and Conservation Center near Wolong.

August 3, 2010 the 8 year old female panda Cao Cao gave birth to a tiny male who weighed 205 grams. This birth was unattended by staff and as close to a natural birth as they could make it.

The first training enclosure was 4,000 square meters in size. The enclosure is located at an elevation of 1800 meters. Cameras were hidden all over this enclosure to allow the panda keeping staff to monitor the pandas without coming in contact.

During his stay in this enclosure Tao Tao learned to climb trees, where he sometimes stayed for as long as 18 or 20 hours in a day. His food was mother's milk.

On December 3, 2010 Tao Tao was moved from the enclosure to a veterinary facility where he was given a physical showing him to be in good health.

Phase 2 began February 20, 2011. Tao Tao and his mother are moved to the first "Field Training Wilderness Enclosure". This is actually two enclosures totaling 42,000 square meters and located 2,200 meters above sea level. This enclosure has more than 50 cameras installed.

The panda staff must all wear panda costumes smeared with panda urine and panda poop when going anywhere they might be seen by pandas undergoing wilderness training. There are some detractors of this costuming who claim that since the pandas recognize people by scent, this is an ineffective method. I think it works even if not perfectly.

During this phase Tao is learning to walk further, he learns to climb trees better and now sounds of wild animals are broadcast to help train him to identify sounds he might hear when released into the wild.

Phase 3, May 3, 2012: The mom and cub are moved to an even larger wilderness enclosure. This time they have 240,000 square meters to roam in at an altitude of 2,000 meters.

Here Tao learns to forage for bamboo. He takes a big fall from a 10 meter high tree, but is unharmed. Sounds are being broadcast to help distinguish dangerous from non-dangerous animals.

In May, 2012 the National Forestry Ministry Protection division held a meeting. Titled **"The summary of Wolong Giant Panda Tao Tao Wild Training and Reintroduction Experts Argumentation Meeting"**. More than 30 Chinese panda experts attended this meeting.

This meeting was a typically Chinese move to help the decision makers save face in the event that Tao Tao meets the same fate as Xiang Xiang. The experts achieved consensus and recommended that Tao Tao be released into the Lizipeng Nature Reserve.

This Lizipeng Nature Reserve is estimated to have a small population of only 15 pandas and is seemingly a wise choice. With lots of bamboo and a low panda density, Tao Tao will have an excellent chance to survive.

Phase four, October 11, 2012: Tao Tao weighing 42 kilos is moved to the Lizipeng Nature Reserve and after 26 months of wilderness training, he is released into the true wilderness.

As I write this the spring mating season is just beginning. To date Tao Tao is surviving just fine. He faces a test now as other males in the area begin searching for female companions to mate with. This competition with other males most likely is what led to Xiang Xiang's death.

A puzzle to me is why the Chinese decided to release a second male panda. After Xiang Xiang died most of the talk going around was that the next panda released would be a female. The thinking was that a female would not have to deal with aggressive male pandas attacking during the mating season.

There are now six female pandas in the semi-wild enclosures at Hetaoping, beginning their artificial wilderness life, phase one in the next captive release effort. Next they will give birth and then plans are to subject their offspring to the same four steps as Tao Tao was put through . China is moving forward with the captive release program. Now we all need to sit back and see just how Tao Tao does in the wild.

Saving the giant pandas from extinction is a huge undertaking. Successfully releasing pandas into the wild can be an asset in gaining genetic diversity within the panda population. But without more and better places for the pandas to live, the ultimate success of the Panda Conservation Program is questionable.

Keep your fingers crossed and together we'll see how Tao Tao does in the coming months and years.

This idyllic farm scene in the Wolong Valley seems peaceful and unthreatening. But look closer and all that's within your view is habitat lost to the pandas and other wildlife. The trees growing on the slopes are young, with no bamboo growing beneath them. A bamboo planting campaign is ongoing in these mountains. Several hundred acres out of tens of thousands of de-forested acreage have been planted with bamboo.

My friend and panda keeper Zhang carrying what I think is the cutest panda I ever saw.

Chapter Five

Dangers to Wild Pandas

Loss of habitat is one of the most serious dangers the panda population faces today. Detractors of the captive breeding and captive release programs claim that by focusing on these breeding and release efforts, the real danger to the panda population is being ignored. They claim that habitat loss continues unabated. Their belief is that there is no focus by the government to stop the progress of civilization into the wilderness areas.

As I've traveled through China, I've looked at the country side and the mountains wondering if the pandas were being given enough space to survive. China is a huge place. I have hiked into the mountains around Wolong and into the Qinling Mountains closer to Xian in search of pandas in the wild. In the mountains around Wolong I spotted two red pandas, but no giant pandas. I did see wild giant pandas in the Qinling Mountains.

While hiking I saw large areas that had almost no human activity. I saw other areas where critical panda habitat was being cut down as the loggers moved through the forest. One man cannot grasp the overall picture of habitat loss versus habitat rehabilitation. The scope is just too large.

In my heart, I believe that China is moving forward by slowly relocating local people from some panda habitat. While in other areas the population of those local people continues to grow. In some areas logging and tree cutting have been stopped. In other areas the logging continues unabated.

One thing is certain. Since 2000 when the Panda Conservation plan first outlined a plan to set aside 35 Panda Nature Reserves, the number of Panda Nature Reserves and Preserves has grown. There are now 60 designated Panda Reserves in China. It's not a question of whether or not the Chinese government is doing something about habitat. Rather it is a question of whether what they are doing will be enough.

Predators can also pose a danger to the wild giant pandas. In the mountains where the pandas live the smallish clouded leopard, weighing around 50 pounds poses a danger to the young pandas, but probably not to a full grown adult.

Asian black bears are roughly the same size as a full grown male panda, but can grow larger, up to 200 kilograms. They present a danger to the smaller giant pandas.

Tigers at one time inhabited some of the same forests as the pandas, but they are gone now. Packs of wild Asiatic dogs have been known to injure and kill giant pandas.

Finally pandas can pose a serious threat to one another. In captivity pandas older than 3 or 4 are generally either kept alone or together with only one other compatible panda. In the wild males will fight for territorial superiority, especially during the mating season. While the pandas generally don't die during these encounters, the wounds can kill them later.

Hunters at one time were the very worst enemy the giant pandas had ever faced. In the 1980s and 1990s the pelt of a giant panda was so valuable they were selling for as much as $50,000 to $100,000.

Very stiff criminal penalties including death and life imprisonment slowed the poaching and illegal trade in giant panda body parts. The national love of pandas by virtually every child and by most adults makes trading in illegally hunted panda parts difficult to hide. It would be a rare child who saw a dead panda in or near his village without his mentioning it to other children and to his teachers at school.

Today it is not unusual for local mountain people to call a nearby panda preserve or forestry office to report that a sick or injured panda has wandered into their village. Every year some pandas are rescued and brought to the panda preserves in this manner.

Pandas live in steep rugged terrain. That is foggy mist in the background, not smog or pollution. We are a hundred miles from any city here. Our hike began on the far side of the mist. It is a wonderfully wild area. The blemishes man has left are slowly fading.

Chapter Six
Tracking Pandas In the Wild

Few people other than scientists, researchers and the local people living in the mountains have ever seen a panda in the wild. When tourists visit China they travel to cultural sightseeing locations rather than to outdoor nature related locales.

I've had the good fortune to have time to hike days on end in the wilderness mountain areas of China in search of panda and other indigenous wildlife. My luck with wildlife helped me find two red pandas, 2005 and 2007 while hiking alone in the forests around Wolong.

I've also seen wild pandas twice, both trips I was accompanied by trackers & guides who helped me get close enough to see the pandas. Their technique was to place me in a strategic position above them on the bamboo covered mountainside. My guides then walked a lower path.

Both times the wild pandas moved uphill away from the trackers and thus crossed my path where I was just sitting and waiting as my hard working guides bulled their way through the dense bamboo.

The pandas were great to see like that. But equally exciting for me was viewing the rare golden monkeys. These primates' favorite food is fungus or mushrooms growing high in the trees.

One of the most dangerous animals in the Chinese forest for humans is not the leopard or bear. It is the Golden Takin., a large moose-like animal that is related to musk ox. The Takin is a powerful animal with a hefty set of antlers.

If frightened a Takin will run fast and furiously along a trail. Any human in the way is just trampled. My guides have always been wary of any chance encounter with Takin.

Other animals that share habitat with giant pandas include the goat like serow and goral.

My hikes in the mountains of China affected me more than any of the hundreds of hikes I have ever done. I have hiked in the mountains, on the beaches and in the deserts of Thailand, Laos, The Philippine Islands, Canada, Mexico and the United States. But the stark contrast between these wild places and the big cities of China is so dramatic, it's as if I was on an alien planet.

Chapter Seven

Tourists and Pandas

Tourism is an important means of bringing large sums of money into the panda breeding program in china. Wherever there are pandas, there are crowds of people who want to see these unique and interesting animals. You decide, is this good or bad for the pandas?

Zoos across the world get pandas on a lease loan program from China. These programs are usually 5 or 10 year plans. China earns roughly one million dollars a year from each loaned out panda. Many people speculate about the high cost to the various zoos and whether or not the pandas bring in more revenue than they cost. That is not a topic I will try to analyze here. But I will say that dozens of zoo administrators around the world have analyzed the likely cost vs. income and concluded they wanted to have pandas housed at their zoo facility.

Sometimes people can become almost compulsively obsessive about the pandas. That's one reason there are a number of live panda cam feeds on the internet.

Veronica is a friend who traveled with me to see the pandas in 2009. She is fascinated with pandas, especially the baby pandas. She works a solitary night shift responding to alarms for a security company. Her work is characterized by long periods of boring waiting time, simply sitting or standing These long periods of patient waiting are broken by brief spells of intense stressful activity as one or more alarms sounds and she reacts appropriately.

She works at a large desk complex with some video monitors and lots of lights, alarm sounds and telephones. Prominently placed on the left and right corners of her work area are two video monitors. During those long, boring night shifts Veronica has those two monitors tuned non-stop to two different live panda cam feeds. She changes which panda cams she watches, but 5 nights a week, those two monitors bring her live pandas to help break up the monotony of her job.

For someone like Veronica, the opportunity to hold a one year old panda in her arms was something beyond imagining. When she learned she could do this in China, Veronica began saving for what would be the most fulfilling dream trip of a lifetime for her. After 4 years she had saved enough to go and see and be close to real pandas.

The Chinese administrators of the captive panda facilities know that people like Veronica will pay almost any price to have a special few minutes with some pandas. They have gradually increased the fees they charge while also adding more and different fee based activities.

In the beginning, at the Beijing Zoo for instance in the 1990s for a small tip of a few dollars a panda keeper would allow you to snap a photo or two with one of the pandas. That tip went directly into the panda keeper's pocket. Gradually the "sitting with a panda" photo opportunity became more official and fees were paid directly to the panda preserve. From $5 the cost has risen so now in 2013 the cost for a photo op at Bifengxia, to sit with a one year old panda for 30 seconds to one minute is about $125 dollars.

Other photo and close contact opportunities are available. At Bifengxia a 3 minute play period with one or more one year old pandas costs about 1,000 to 1,500 rmb or $150 to $250 dollars. The fees are going up rapidly and by 2014 I would not be surprised if these are fees are doubled.

A very special and tightly controlled situation that is offered occasionally at Bifengxia when the panda babies are just the right age, is a one or two minute time period inside the baby nursery to bottle feed and or burp a young several month old baby panda. The cost for this really special encounter started 4 years ago at 5000 rmb or about $800 dollars. Last year I heard they offered this for about one month and the cost was more than $2,000 for the 2 minutes in the nursery,

You might think that only a crazy person would spend $2,000 for 2 minutes, but when this opportunity is offered, there are tourists paying the price every day to have this totally unique chance.

My point of mentioning these fees is to show that really large sums of money can be generated by creative, highly capitalistic, exploitation of the captive breeding panda population. Is this right or wrong? I don't have the mental makeup to make a judgment. I've given you some of the facts, you can form your own opinion.

I do know it would be almost inconceivable for a zoo in the United States to sell 2 minute play periods with a baby panda for – oh say $500 per two minute session. Even though if offered there would probably be a line from morning to closing of people waiting their turn to pay $500 for a 2 minute play time.

My tour groups go to the Bifengxia Preserve where they do several days of volunteering work alongside the panda keepers. Volunteering really does give the panda keepers help with their chores. Tasks such as cleaning up hundreds of pounds of panda poop are easier with a helping hand. The volunteers make life easier around the panda preserve and probably allow for a slightly smaller staff size.

But what volunteering really does is to bring substantial sums of money directly to the panda preserves. Money those preserves use directly for staff and for new and ongoing projects at that particular facility.

Panda bites the thumb that feeds it

There are detractors of the volunteering panda program. Those detractors say the possibility of disease transference is increased when more people come in contact with the pandas. Others say it can be dangerous for the volunteers, who are usually inexperienced at working with wild animals.

The program has been running now since 2004, approximately 9 years. In those 9 years thousands of volunteers have assisted the panda keepers at various panda preserves around China. My tour volunteering groups have done volunteer duties at Wolong, Bifengxia and Lougantai panda facilities.

In all this time there has only been one serious injury to a volunteer reported. There has not been any case of known transmission of disease from a volunteer to a panda. These are really excellent and strong statistics validating the safety for both the volunteers and the pandas.

I know that now you're wondering what that one injury to a volunteer was. Here's the story as I heard it soon after the incident occurred.

A woman only ever identified as an American, aged 50, named Lisa was performing short time volunteer duties when she was injured. Those of you who have gone with me know you are given some simple safety instructions. Included is a warning not to go near the bars of the cages nor to feed the panda unless directed to do so by the panda keeper.

When hand feeding the pandas in their cage we generally give them panda bread, apple or sometimes carrots. I have never known a keeper to allow hand feeding of small pieces of bamboo to the pandas.

The story as I heard it is that while the panda keeper had her back turned, Lisa the volunteer took some short pieces of bamboo and attempted to feed one piece to the panda.

The panda reached out to grasp the bamboo and also (probably by accident) grabbed hold of Lisa's hand or the glove she was wearing. The panda then bit down on the bamboo stalk, simultaneously biting the volunteer's thumb.

The surprised woman tried to pull her hand away, got excited and was screaming all of which caused the panda to bite down harder, chopping off the end of her thumb.

According to a local Wolong doctor the woman lost 20% of the end of her thumb.

I am very pleased to say none of my volunteers has ever done anything like that. Nor have any of them been bitten or clawed by a panda.

So the question immediately asked by American internet trouble makers was should the program be stopped because of this one injury?

My reply to this is of course not. There is no activity that humans engage in that doesn't result in injuries. I was surprised the first time I heard how many people fall in their bath or shower every year. According to the National Safety Council, one person dies every day from using a bathtub or shower in the United States. Additionally 370 people are injured in bathtub or shower accidents every day.

One tiny portion of a thumb doesn't seem such a large injury rate. At the time the accident happened in October of 2006 1,200 volunteers had performed volunteering activities at the Wolong Panda Preserve.

I'm a big fan of the volunteer program. I think the money generated by the volunteer fees, plus the many donations that volunteers make while at the panda preserves are a big help to the budgets of the preserves. And on a more personal note, I was once a celebrity because of volunteering!

It was 2006 when I became a minor celebrity and had my 5 days of fame at the Wolong Panda Preserve. This came about because of my leading volunteering groups.

In April I brought a group of 11 to Wolong to do volunteering work. While there the panda club photographer took a group photo of me with my group of volunteers. They were all clad in the drab brown full length coveralls that the volunteers wear. I was wearing my normal black fleece jacket.

Later in the year and unknown to me before it happened, the panda preserve used that photo to make a sign that told visitors about the volunteer program at Wolong. Our group photo was enlarged and placed prominently in the center of the sign. The sign was installed just after the main entrance gates, near the panda hospital. Virtually every Preserve visitor would see that sign as they entered the Panda Preserve.

One day morning around 10:00 a.m. I was sitting on a concrete bench along the main walkway, near where there are several panda enclosures. My volunteer group members would pass by this bench as they went to get fresh bamboo or to take the day old bamboo to the bamboo grinding and disposal area. Sitting here it was easy for them to ask me questions if they had any. It was also normally a quiet location with shade or sun depending on which bench I sat upon.

That morning as the first bus of tourists arrived, I noticed some of the Japanese tourists taking photos of where I was sitting. I didn't think anything about it because it is common for me to have Asians take my photo. As the various members of this particular group passed by me, they would stop, then take my photo.

After 30 minutes or so a group of teenagers passed by. One of them stopped, looked over at me, then called excitedly to the others in the group. Before I knew exactly what was happening there were teen kids sitting on both sides of me on the bench.

There were kids giggling and laughing, standing behind me and posing with their fingers in a V or W. I asked what they were doing and one girl who spoke some English said something that sounded to me like 'photo with famous guy". But she was laughing so hard, I wasn't sure exactly what she said.

An hour or so later another bus arrived and the same thing happened all over again. I gave up on writing in my notebook and set out to find the cause of these photo taking frenzies.

I had to laugh as I walked up to the sign where the workers were still filling in around the signposts with some concrete. My photo, along with that volunteer group was what had caused all the commotion.

This photo taking went on for 3 days and I finally grew weary of it. On the last day I arrived very early to the Panda Preserve, entering as the guards first unlocked the gates to allow the workers entry. Being known by everyone, I was not questioned as I entered the Preserve an hour before anyone else would arrive. Casually I picket up a small glop of mud and then very carefully blurred my face on that sign. Standing back a couple feet, I was no longer recognizable in the photo, but the sign looked unblemished. So ended my one week of celebrity.

About the Author

Here's the tweet version.

Ex-hippie vagabond, lives in bamboo hut on a small tropical island & has a passion for living life fully. He tries to live every day as if it were his last. Really!

I traveled to China for the first time in 2004 to develop a tour designed to bring travelers to the Wolong Panda Preserve to work as volunteers. I wanted to give my clients a unique experience impossible to have anywhere else in the world.

Nine years later, I have many fond memories from the panda preserves and the other places I've traveled to in China.

Travel and getting close to wild animals is what my life is about. I have led whale watching tours in Mexico for twenty years. The migration of the gray whales has been the metronome that set the beat for my entire life.

The panda tours I lead are not tied to a migration rather to the weather. I find it is just too cold to be in the China mountains before April or after November.

I have written two previous books that are available on Amazon.com. **Ghost Tripping** is a compilation of short travel stories, each story with some ghostly element about it. **Gray Whales My Twenty Years Of Discovery** is a book about gray whales.

When I started writing this book, I was actually working on a book about blue whales, but the pandas overpowered my mind and made me write this book. So good or bad, whether you like the book or not, I blame it all on the Giant Pandas.

Books I've written and available on Amazon.com
Gray Whales My Twenty Years of Discovery

Ghost Tripping

I welcome comments or conversation with you, my loyal readers. Yes, I still lead tours to China, Thailand, Mexico, the Arctic, and Africa. You can join me if you want.

How to contact me:

Email is best. If you don't get a reply within 48 hours I did not receive your email or I'm on a wilderness trip.
keith@greywhale.com or
rowman1998@yahoo.com

You are welcome to chat with me at any of the following venues if you find me online. Just add me to your chat list.

Yahoo messenger ID = **rowman1998@yahoo.com**

Skype ID = **bajajones**

My tour business website is:
www.bajajonesadventures.com

If you enjoyed this book I ALWAYS appreciate when my readers write a nice review and post it on Amazon.com along with – hopefully - one of those nice 5 star ratings.

Coming December, 2013
Blue Whales: Ten Years Of Discovery

I'll leave you with these thoughts:

- Is tourism good or bad for the pandas?

- Is exploitation of the pandas, to generate income at zoos, good or bad for the pandas?

- Is it ethically and morally acceptable to exploit several generations of captive pandas in order to save future generations from extinction?

- Although there is no recorded instance of pandas getting sick from contact with tourists, is this contact good or bad and should it be allowed?

- Should the captive breeding program be ended or reduced in size and scope?

- Should captive born pandas be placed at risk by being released into the wild?

- Does pandamania detract from other serious wildlife conservation programs?

- Should conservationists stop publicizing their efforts to raise money to save the pandas in China and instead say they are raising money to save the endangered bamboo rat?

The End

Suggested Panda Reading

There are a couple books I will suggest, but truthfully the panda book market is flooded with books aimed at children or else written by some scientist whose writing style is guaranteed to put the reader to sleep in 5 minutes.

Dr. Sarah Bexell, a Director level researcher for about 10 years at the Chengdu Panda Base (also holds concurrent positions in USA) and along with **Dr Zhang Zhihe** the Chinese Director of the Chengdu Panda Base wrote. I recommend this book. Although I might not agree with all her thoughts, she is far more educated in these matters than I am:
Giant Pandas: Born Survivors

Dr. George B. Schaller who was one of the first American researchers in China's wilderness in the 1980s wrote this somewhat downbeat book in 1990. But a lot has changed since then. Dr. Schaller is still active in Panda matters. I think he and I don't agree about the benefits of tourism and pandas.
"The Last Panda"

Chris Catton wrote the book long considered the best overall popularly written book about Giant Pandas. First published in 1990, the book is definitely dated now.
"Pandas"

Michael Kiefer
This is a story about the first captive panda brought to the United States. I didn't enjoy this tale, but many people have liked it a lot.
"Chasing the Panda"

I'm sure each of you panda crazy readers has several favorite panda books. I hope my book becomes one of those favorites.

www.ingramcontent.com/pod-product-compliance
Lightning Source LLC
Chambersburg PA
CBHW040842180526
45159CB00001B/279